LOOKING AT SCIENCE 1

A First Look

David Fielding

Basil Blackwell

© 1984 David Fielding

All rights reserved. No part of this publication may be reproduced, stored in a retrieval system, or transmitted in any form or by any means, electronic, mechanical, photocopying, recording or otherwise, without prior permission of Basil Blackwell Publisher Limited.

First published 1984

Published by Basil Blackwell Limited
108 Cowley Road
Oxford OX4 1JF

ISBN 0 631 91350 5

Printed in Hong Kong

Topic symbols

◆ This work is about air and water.

◆ This work is about animal life.

◆ This work is about electricity and magnetism.

◆ This work is about light and dark.

◆ This work is about mechanics.

◆ This work is about plant life.

◆ This work is about weather and climate.

Look for the symbols in the other books in the series. There is more work about these things in the other books.

Contents

A word to teachers and parents   5

What is science?   6
How science has changed human life. What science is and how scientists work.

Part 1   Our world

Alive and not alive   8
The world contains animals and plants, and things with no life at all. How animals and plants are alike, and how they are different.

Growing things   10
What plants are like and how they live. The variety of plants.

Under your feet   12
The animals in the soil, how they live, and why the world needs them.

Our world   14
What our world is made of. How all its different parts belong together.

Part 2   Earth, moon and sun

Invisible but real   16
Our world is surrounded by air. Air cannot be seen, but is powerful.

Night and day   18
What makes night and day happen, and why they follow each other.

Light and shade   20
Why we have shadows in the daytime. Why they change shape and direction during the day.

Earth, moon and sun   22
Stars, planets, moons and our world travel through space.

Part 3   Things we cannot see

A powerful pull   24
Why things fall towards the ground. What makes things heavy. A simple way to measure heaviness.

The rough and the smooth   26
What friction is, and how it makes it hard to pull things along.

Unseen attraction   28
What magnets are and what they do.

Things we cannot see   30
We cannot see gravity, friction and magnetism. People have been exploring these forces for centuries.

New words   32

Acknowledgements

Janet and Colin Bord 14(2), 21(3)
Central Office of Information 13(3)
London Transport 27(3)
Milk Marketing Board 9(2)
The Northern Picture Library 23(3)
Oxford Mail and Times 25(2)
Anne and James Robertson 16 (1, 2)
Rebecca Skillman 13(4), 21(4), 29(3)
David Waterman 30(1), cover
ZEFA 14(1), 22(2), 25(3)

Illustrations by Michael Stringer (colour)
and David Fielding (black and white)
Design by Indent, Reading

A word to teachers and parents

A First Look is one of five books in the *Looking at Science* series. The series has been designed to do two things:

- It gives children a solid body of knowledge in natural and physical science.
- It begins to introduce the nature of scientific enquiry.

These two elements are developed side by side through the books.

Each double page covers a particular area for study. The left-hand page outlines an activity to perform, while the right hand page gives information connected with it. The activities are introduced with the symbol ❤, and cover **experimentation, observation and recording**. A list of all the equipment needed for the experiments is given near the beginning of each spread.

Each book also contains another kind of double page, which is purely factual, spaced at regular intervals. These pages draw together the themes of the preceding pages.

These pages can be worked through in order. Alternatively, they can be used as source material for topic work. Suggested topic areas are identified, with symbols, in the contents list.

Notes for Book 1

A First Look is the introductory book to the series. It presents, simply, themes that are developed in detail in the later books. At the same time, it is a self-contained book in its own right, and can be used without the others.

It begins by explaining, on pages six and seven, how science has changed human lives, and what scientific enquiry is. From then on the book falls into three parts:

Part 1   Our world
Part 1 looks at natural science. It explains how the world contains inorganic minerals such as stones and soil, and organic things such as plants and animals. It looks at the differences between minerals, plants and animals. It takes a separate look at plants, and begins to outline some distinguishing characteristics of plants.

It explores the simple types of animal life that live in soil. It describes some of the characteristics and functions of these soil animals. Part 1 ends with a general description of the world. This description explains how organic and inorganic things all fit together into a working system that we call 'the world'.

Part 2   Earth, moon and sun
Part 2 moves towards physical science. It shows that the air surrounding the world has real substance, despite being invisible. It explains how the turning of the world causes day and night to happen. It goes on to explain how our daytime shadows are made and why they change shape and direction as the day passes. The concluding section describes how the sun, Earth and moon float in space, and relate to each other. It outlines the nature of stars, planets and moons.

Part 3   Things we cannot see
Part 3 deals purely with physical science. It explains what gravity is, and how gravity causes things to have weight. It looks at how friction makes it hard to move things, and how it can be lessened. It explains what a magnet is, and what effects magnetism has. The final section draws these ideas together by explaining how the world contains invisible forces; how men have learned about them; and how men have learned to use them.

What is science?

People before science began

Long ago, people could not make clothes. They wore animal fur instead. They could not make fires, so they could not cook food. They did not know how to grow food. They had no medicines. They could not build houses or boats. They did not know how to make wheels for carts. They could not make metal, and so they had no machines.

How science has changed us

Modern people have warm, woven clothes. We make heat and light from coal, gas, oil and electricity. We know how to grow plants and keep animals. We know how to turn them into food. We have medicines for when we are ill. We know how to make strong, warm buildings. We have machines to make our lives easy, and cars to carry us around.

What science is

People changed their lives by finding out why things happen. Then they found out how to make things happen when they wanted. For example, they saw that plants grew. They worked out that the plants grew from seeds in the soil. Then they learned how to plant the seeds. They learned to grow the plants they wanted.

This way of finding things out and trying out ideas is called *science*. A person who does a lot of finding out is called a *scientist*.

Picture 1   This shows how early man lived and how some people live now. Can you spot five differences between the pictures?

Picture 2   This is how a town might look after an earthquake

How a scientist works

Imagine a town after an *earthquake*. Some walls have fallen down completely, but some have lasted better. A scientist notices this. He wonders why some walls were stronger. He looks for clues. He wonders if the way the bricks were arranged made the difference.

He decides to *experiment* (you can try this experiment yourselves). He builds model walls. He arranges the bricks differently in each wall. Then he makes 'earthquakes' by shaking the walls. He records what happens by writing down how well each wall lasted. He draws them, too. Then he thinks about what he has discovered. He decides that walls are stronger when the bricks overlap a lot. He works out the strongest way to build a wall. He tells people how to build stronger walls in future.

Science and this book

This book gives you experiments to try and ideas to think about. It aims to help you to think like scientists. It tells you interesting things that scientists have already discovered.

Picture 3   Test different ways of building walls

Part 1   Our World

Alive and not alive

The world is full of surprising differences. Even a little bit of soil can be full of things that are different from each other.

 You will need a bowl, a trowel, newspaper

♥ Experiment: Investigating soil

Find an area of weeds. Dig up some of the weeds, with plenty of the soil under them. Put the weeds and soil in a bowl, and bring them inside. Copy this chart into your book:

Not alive	Things that do not grow or move
Alive	Things that grow in one place
	Things that move around by themselves

Now carefully spread your weeds and soil on a sheet of newspaper. Look through the soil and see what things you can find in it. Write on your chart everything that you find. Write it on the line where you think it belongs. When you think that you have found everything, put the soil back in the bowl. Tip it back where you found it.

Picture 1   The experiment should look like this

 Record

1   Describe how you got your soil. Say how you searched through it, and what you found.
2   Say if the things you found were unlike each other. Write down what the main differences were.
3   Give a name for things that grow in one place.
4   Give a name for things that move around by themselves.

Organic and inorganic things.

There are two main types of things in the world: things that are alive and things that are not alive. You will have found some in your experiment. It is usually easy to tell which is which. Things that are alive need to feed. They grow, and later on they die. They make new copies of themselves before they die. Things that are not alive cannot do any of this.

Things that are alive, or used to be alive, are called *organic*. Things that are not alive, and never were alive, are called *inorganic*.

Inorganic things are important, even though they are not alive. The world is mostly made of inorganic things. The world is mostly made of rock, and rock is inorganic. The air we breathe is inorganic. So are water and soil.

Organic things, because they are alive, are different from inorganic things. But not all organic things are alike. There are two main types of organic things: plants and animals. One of the main differences between them is that animals can move themselves around. Plants cannot do this.

Picture 2
Organic things need food and water to stay alive

Picture 3
Plants are organic. New ones grow from seeds

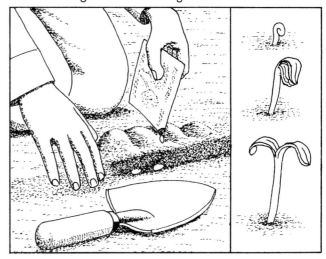

❤ To write

1 Name four organic and four inorganic things which you can see in Picture 4.
2 What are the two main kinds of organic things?
3 How is a dog different from a daffodil, apart from its shape?

Picture 4   The world is made up of organic and inorganic things. How many of each can you see in this picture?

Growing things

The world has many different kinds of plants. You are going to search for some of them.

 You will need plastic bags, a trowel, a metre ruler, a 30cm ruler

First, you need to make this chart:

Height of plant	Type of plant
over one metre	
between 30cm and 1m	
between 5cm and 30cm	
under 5cm	

Go to the area your teacher shows you. Use rulers to find plants over one metre high. Write on the chart what kind of plants they are. Take a sample leaf from each plant and put them into a bag.

Next, find plants over thirty centimetres high but less than one metre. Take a sample leaf from each plant and put them into a second bag.

Carry on until you have finished your chart. When you find small plants, dig one up to put in a bag as a sample. (Never do this outside school. Some wild plants are dying out. Too many are being picked.)

Picture 1   Collect your plants carefully and find out the name of each one

♥ Record

1 Describe your search. Say how many different kinds of plant you found.
2 Find the right name for each of your samples. Ask your teacher to help you.
3 Make a list of how your plants are alike, and how they are unalike. Think of their colour, shape, softness, size and smell.
4 You have sorted plants into different sizes. How else could you sort them?
5 Describe what a plant is, to someone who has never seen one.

What is a plant?

A plant is a living thing. This means that it grows and changes. In the end it dies; but before it dies it makes new copies of itself. Because it lives and grows, it is called an *organic* thing. Stones, soil and water are not alive. They are called *inorganic* things. Another word for them is *minerals*. Plants need inorganic things. A plant cannot live without soil, water, air and sunlight.

Types of plant

There are thousands of different kinds of plant. Here are some of them.

Most plants are alike in some important ways. They have a special green stuff in them. This green stuff lets a plant turn sunlight, air and water into a kind of food for itself. Most plants have leaves to 'catch' sunlight and air. Most plants have roots to reach water in the soil. Most plants have flowers, though they are sometimes small.

♥ To write

1 How are organic things different from inorganic things?
2 Plants cannot live without ———, ——— and ———. (Fill the gaps.)
3 Can you say why plants have roots?

Picture 3   Name the different parts of this plant

Picture 2   Some of the plants you may see in your search

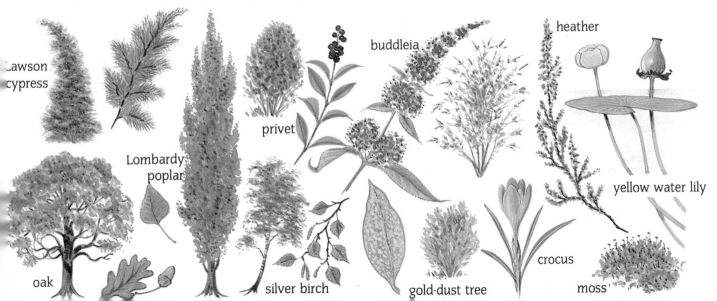

Under your feet

Picture 1 Some soil creatures to look out for

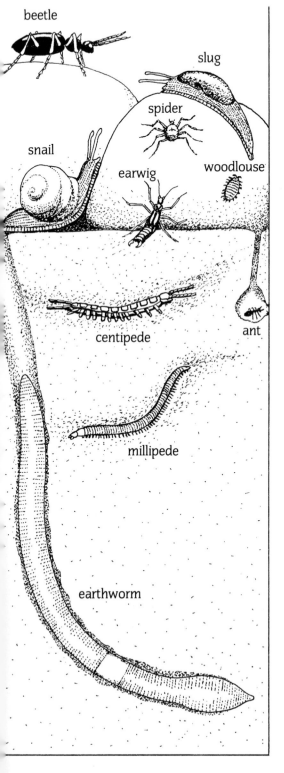

Many little animals live in the soil. See if you can discover some.

 You will need a trowel, a bowl, newspaper, a magnifying glass

♥ *Experiment: What animals live in the soil*

First, you will need to make this chart:

Animals that we found in the soil	
beetle	snail
earwig	spider
millipede	slug
woodlouse	ant
worm	centipede

Dig some soil from under some weeds. Spread this soil out on newspaper, bit by bit. If you find any animals, see if they look like any of those in Picture 1. Tick the chart in the right place for each one you see. Use a magnifying glass to help you see them clearly.

When you have searched through all the soil, put it back from where you got it. Be careful not to hurt any of the animals.

Picture 2 This experiment must be done carefully so that you do not harm the creatures

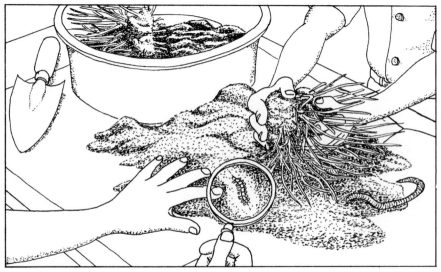

♥ Record

1   Describe your search. Describe any animals that you found and try to draw them. (Do not copy the pictures in this book.)
2   Make a list of ways in which these animals are alike. Make a list of ways in which they are unalike. Think of their sizes, shapes, colours and the way that they move.
3   Say if you think these little animals are good or bad for the soil. Write down any way that they might help it.

Soil animals are useful

Most of these creatures help us. They help get rid of dead things. Some of them eat dead plants. Earthworms, woodlice, snails and millipedes do this. Some eat the bodies of dead animals. Beetles, slugs and earwigs do this.

They also help the soil. When they have eaten, waste comes out of their bodies. This waste is good for the soil. It makes it better for plants to grow in. Worms make the soil looser by tunnelling through it. This lets air get into the soil. Air is good for the things that live in the soil. Loose soil helps rainwater sink to the roots of plants instead of lying in puddles.

Why they live underground

These little animals need to be safe. Larger animals will eat them if they can. They hide under stones and in the ground.

Most of these little animals need to keep their bodies damp. If they get too dry, they die. That is why they live in dark, damp places.

♥ To write

1   What might happen to the soil if there were no soil animals?
2   Why is it good for soil to be loose?
3   Why do worms go deep into the soil in summer?

Picture 3   Things do not grow on hard, poor soil and water cannot drain away

Picture 4
Good, loose soil helps plants to grow

Our world

Picture 1 The planet Earth is a huge ball

Picture 2 The Fen lands look absolutely flat

Planet Earth

We live on a huge ball, or *globe*, called a *planet*. The name of our planet is Earth. This ball is so big that we cannot see how round it is. The Earth seems mostly flat to us, because we are seeing it from low down. People who go high up, in aeroplanes or spaceships, can see that it is round.

Air and moisture

The world is covered by air and clouds. We can see the clouds easily when they are in the sky. We cannot see the air, but we can feel it. We feel the air blowing past us when there is a wind. We can watch birds flying through the air. Birds flap their wings against the air in order to move.

We need the air, because we breathe it. Animals and plants need the air. We need the clouds, because rain comes from the clouds.

Plants and animals

There are many animals and plants in the world. Some are large and easy to see, such as elephants and trees. Some are tiny, such as soil animals and weeds.

Animals are different from plants because they can move around. Animals need to eat things to live, but plants get food straight from the sun and the soil.

Although animals and plants are so different, they need each other. Soil animals look after the soil. They help to make it rich and loose. This helps plants to grow. In turn, plants provide food for animals.

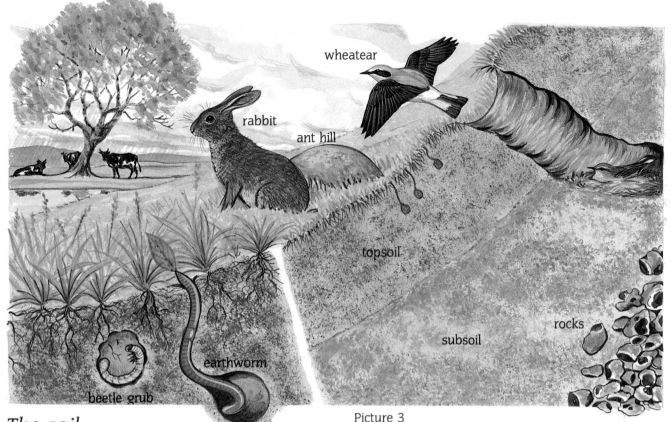

The soil

Most plants grow in the soil. The soil can be as deep as several metres in some places. In other places, it may be only a few centimetres deep. The most important part of the soil is the top part. This is called the *topsoil*, and is only ten or twenty centimetres thick. This is where most plants have their roots, and where most soil animals live.

The soil beneath the topsoil is called the *subsoil*. This does not have much food for plants, and is mostly empty of animals. If the topsoil is taken away, the subsoil cannot grow strong plants.

Rock

Beneath the subsoil is cold, hard rock. This is the rock of the *Earth's crust*. The crust is many kilometres thick.

Below the Earth's crust, far under our feet, the Earth is made of hot rock. Very deep down the rock is *molten*. Molten rock is called *magma*.

Picture 3

Picture 4 How the Earth is made up

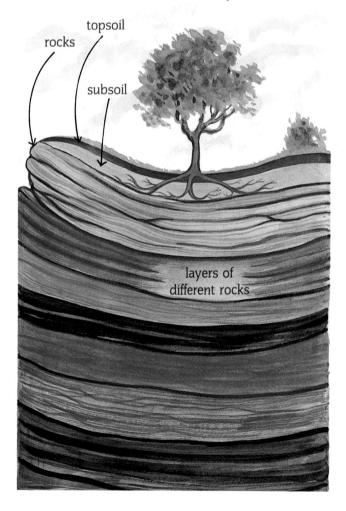

Part 2 Earth, moon and sun

Invisible but real

Picture 1

Picture 2

The people above are warm and relaxed. The ones in Picture 2 are uncomfortable and cold. What makes the difference? It is the air around them. Air can be strong.

 Two experiments: To show the strength of air

 You will need a bowl, a clear jar, a tube, a piece of cloth, water

Push some cloth to the bottom of a clear jar. Fill a bowl with water. Lower the jar into the water, upside down. The cloth must go below the surface. Wait for a moment, and then lift the jar out. Take out the cloth and feel it. It will be dry.

Fill the jar with water and hold it upside-down in the bowl. Put a tube up inside it, and blow into the tube. See how air bubbles out from the tube into the jar.

Picture 3   Carry out the experiment this way

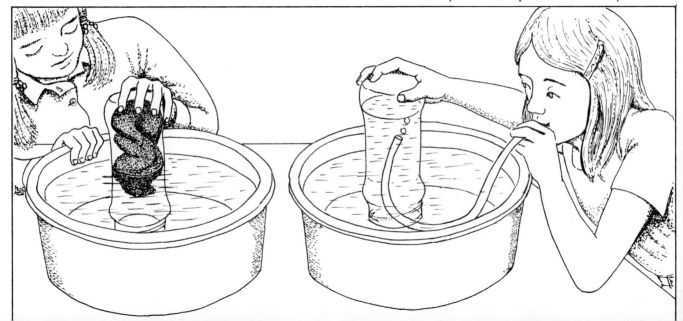

Keep blowing. See how the air pushes all the water out of the jar. See how it tries to lift the jar, too.

❤ Record

1 Write about your experiments, and draw them.
2 Say what they have taught you about air. Is air strong enough to keep out water? Is air lighter than water? Can air lift things?
3 Imagine what you are writing to someone who does not know what air is. Describe what air is like.

Facts about air

Your first experiment showed that 'empty' things are really full of air. Your jar was full of air. This air stopped water from going up inside. It kept the water away from the cloth. Air is strong enough to keep out water.

The second experiment showed that air is lighter than water. You saw this when you blew air bubbles up into the jar. It also showed that air can be powerful enough to push out water. It can be powerful enough to lift things, too. It lifted your jar in the water.

We talk about 'thin air', but air is not just emptiness. Air is something that is all around us.

The atmosphere

Air surrounds the world in a layer. We call this layer the *atmosphere*. The layer of air is thin, compared with the size of the world. But men are so tiny that the layer of air is thick, compared with us.

Many kilometres above our heads, air fades away into empty *space*.

❤ To write

1 Why did the cloth stay dry inside the jar?
2 What is the atmosphere?
3 Why do astronauts need special equipment to help them breathe? Who else needs special equipment for the same reason?

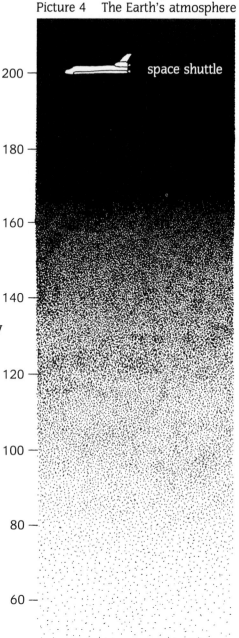

Picture 4   The Earth's atmosphere

Night and day

Why do we have days and nights? Here is an experiment to show you.

You will need a torch, a ball, string, sticky tape, chalk

♥ Experiment: To show what makes day and night

Take a torch to be the sun. Take a ball to be the Earth. Hang the ball on a string. (Use tape to hold it to the string.) Find a dark place and then shine the torch at the ball. This is like the sun shining on the Earth. See how half of the ball is light and half is dark. You have made 'day' and 'night'. Next, make a small dot on the ball with chalk. Imagine that this marks your home. Put your home in the torchlight. Next, turn the ball so that your home is in shadow. See if you can make day and night happen in turn to your home.

♥ Record

1 Describe your experiment, and draw it.
2 Explain how you gave night to your home. Did you move the torch around the ball? Did you turn off the torch? Or did you turn the ball?
3 Say how you made day and night happen in turn. Did you turn the torch on and off? Or did you spin the ball?
4 Say how you think day and night happen to the real Earth.

Picture 1 An experiment to show how day and night happen

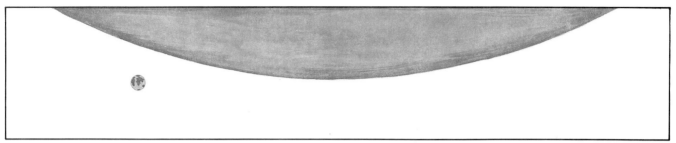

Picture 2 The Earth is very small compared to the sun

Sun and Earth

The sun and Earth float in space. The sun is huge, but very far away.

Imagine the Earth right next to the sun. Picture 2 shows how they would look.

The sun does not go round the Earth

The Sun never goes out. People used to think the sun goes round the Earth, because the Sun seems to move across the sky. Really, it is the Earth which moves. The Earth turns around, just like you turned your ball around in the torchlight. It is the Earth turning which makes the sun seem to move.

Day and night

Imagine a spaceman looking at the Earth and the sun. He sees that half of the Earth is in daylight. Half of it is in night. As he keeps looking, he sees that the Earth is slowly turning.

Countries gradually turn out of night into day. The Earth takes twenty-four hours to turn round once.

♥ *To write*

1 What might happen to the Earth if the sun was closer?
2 Why does the sun seem to move across the sky?
3 From breakfast time Monday to breakfast time Wednesday, the Earth spins round ——— times. (Fill the gap.)

Picture 3 How countries turn from day into night

Light and shade

Why do shadows change shape and direction during the day? Find out with these experiments.

You will need a torch, a toy soldier or doll, paper, scissors, tissue paper or cloth, pencil, card, paper fastener

♥ Experiment: Changing the shadow by changing the light

Find a dark place. Shine a torch at a toy soldier or doll. See how this makes a shadow. Now move the torch from side to side. Move it forwards and backwards. See if the shadow moves or changes shape. Dim the torch with tissue paper or cloth. See if the shadow gets stronger or weaker.

Picture 1   Investigating shadows

♥ Experiment: Moving the shadow without moving the light

Cut a large paper disc. Fix it to a sheet of card so that it can turn. Put it in front of the torch. Put the soldier on the edge of the disc. Draw round its shadow. Slowly turn the disc. Watch the shadow. If it moves from where you drew it, draw it again in its new place. Keep turning the disc. See if you have to keep drawing the shadow again.

♥ Record

1   Describe and draw the experiments
2   Say what makes shadows. Why are there no shadows in complete darkness?
3   Do shadows move when the light that makes them moves?
4   What happens to shadows when the light gets dimmer?

Picture 2   How to make the shadow move

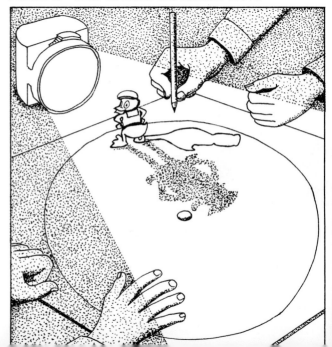

Picture 3 What time is it on the sundial?

Picture 4 Shadows

Where our shadows come from

Shadows are made when light shines past a thing. Where the light is blocked, there is a shadow. Strong light makes strong shadows. On sunny days, shadows are clear. Weak light makes weak shadows. On dull days, sunlight is dimmed by clouds. Shadows are harder to see. You showed this when you dimmed your torch. At night, we have no shadows. There is no sunlight to make them. When we move into the light of street lamps or houses, our shadows are made again.

Changing shadows

If the light moves, or we do, our shadows change shape and direction.

Our daytime shadows are made by the sun. They change during the day. They change length and direction. This is not because the sun moves. The sun seems to move across the sky, but it does not really move at all. The sun stays still. It is the Earth which moves. The Earth turns round in front of the sun. It turns rather like the disc in front of the torch.

❤ To write

1 What makes our daytime shadows?
2 A person stands by a road at night. A car passes, with lights on, from right to left in front of the person. Does his shadow move in the same direction as the car, or in the opposite direction?
3 Why are shadows not clear on cloudy days?

Earth, moon and sun

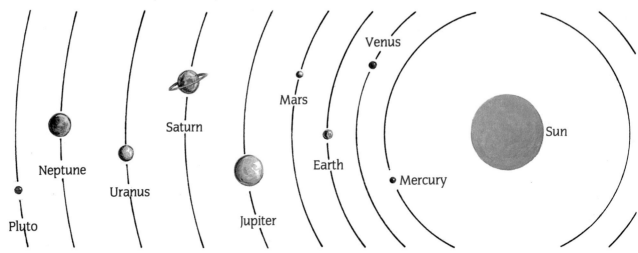

Picture 1   The solar system – the sun and nine planets

The Earth

The Earth is a huge globe, or ball. It measures nearly thirteen thousand kilometres across. If you could walk straight in one direction for years, across mountains and oceans, you would get back to where you started from. The world is so large that it would seem flat to you all the time.

Planets in space

The Earth floats in space. Space is an empty blackness that goes on for ever. The Earth is a planet. There are other planets in space. They too are huge balls of rock floating in space. We do not know of one with air, plants and animals like our planet.

Picture 2   Saturn is one of the planets in Earth's solar system

Picture 3   The sun has burnt this child

Sunlight and air

The Earth is lit by the sun. The light from the sun is strong. You know how badly sunburned you can get on a hot day. The air is helpful to us, because sunlight has to pass through it before it reaches the ground. The air blocks out some of its fierceness. Sunlight has things in it which could damage us, if the air did not protect us.

Stars

The Sun is a *star*. A star is a huge burning ball of gas which floats in space. The sun seems small because it is so far away, but it is really very big indeed. Space contains millions of stars, but because they are so far away they seem small and weak — just points of light. We can see them only at night.

The Earth travels round the sun. It takes a year to go round once.

The moon

A smaller ball of rock circles round the Earth. This is the *moon*. It is not a planet. Planets travel round stars. Moons travel round planets. Some planets have many moons going round them.

Day and night

The Earth and the moon are lit by the sun. The parts facing the sun are in daylight. The parts facing away from the sun are not lit. These parts are in night. If the Earth did not spin, one side would always have day and the other side would always have night. But the Earth spins round once every twenty-four hours. This makes places on Earth move through day and night in turn.

The Moon spins more slowly than the Earth. A day on the Moon lasts two Earth-weeks.

Picture 4   The Earth *orbits* (goes round) the sun annually. The moon orbits the Earth once every 28 days

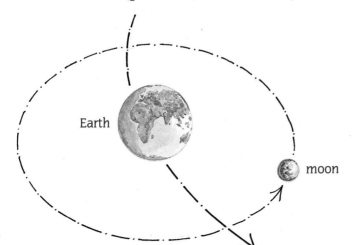

Part 3 Things we cannot see

A powerful pull

Things fall easily. It takes no effort. Yet things do not rise unless they are made to. This takes effort. Why is this?

 You will need strong elastic, weights, objects to dangle, a ruler

 *Experiment: To show how much things want to fall*

Take a one-kilogram weight. Hold it and feel the effort you need to keep it from falling.

Picture 1   This experiment helps you test why things want to fall

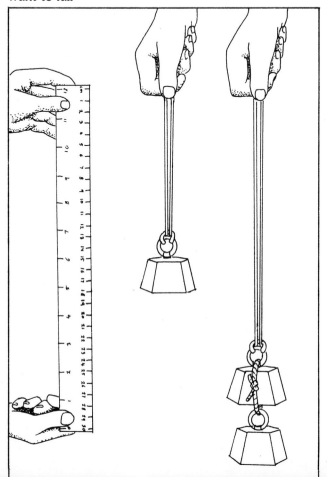

You can see this effort if you use elastic. Fix a strong elastic band to the weight. Measure the length of the elastic while it is slack. Then dangle the weight on the elastic. See how the elastic stretches. Measure it. The amount of stretch shows the effort it takes to hold the weight.

Take another weight, and tie it to the first one. See how far the elastic stretches. (Do not let the elastic stretch until it seems ready to break.) The effort of holding the weight has increased.

You can use elastic bands to hold other things. Try dangling different things, such as polystyrene and Plasticine. See how long the stretched elastic is each time.

♥ *Record*

1 Describe and draw what you have done. Copy this chart into your book, and write in your results.

What we dangled	The length of the elastic

2 Say if it always takes some effort to hold things in the air. Does everything need the same effort?

3 Does bigness always mean heaviness? Are big things always harder to hold up?

4 What makes it hard to hold things up? Where does the pull downwards come from?

Picture 2 Gravity helps to pull down a factory chimney

Gravity

The Earth pulls things downwards to itself. This pull is called *gravity*. Gravity makes things hard to lift. Gravity pulls at everything. The Earth makes gravity because it is so big and solid. We say that it has a great *mass*. It is *massive*. Gravity is made wherever there is a great mass.

Gravity on other planets

Other planets make gravity, too. So does the sun, and so does the moon. Spacemen on the moon feel gravity. It is weaker than the gravity on Earth, because the moon is less massive. Spacemen on the moon can jump high, easily.

Zero gravity

Far out in space, there is no gravity. There is nothing massive to make it. We say there is *zero gravity* or *weightlessness*. In space, things float. They have no weight.

❤ To write

1 Why would zero gravity make life difficult in the classroom?
2 What causes gravity?
3 What would you have to do in order to play football on the moon?

Picture 3 Why do astronauts float in space?

The rough and the smooth

Picture 1   It is difficult to pull things over rough surfaces. On smooth surfaces things move more easily

These men are dragging a heavy log. At first, the log is hard to move. Later, the men find the work easier. What makes the difference?

> You will need a container with a flat bottom, two 1-kg weights, a strong elastic band, masking tape, sandpaper, various surfaces

❤ Experiment: To show the effort of dragging things

See what happens when you pull something. Take a container with a flat bottom. Put it on the desk and tie an elastic band to it. Put two one-kilogram weights into the container. Measure the elastic while it is slack. Now use the elastic to pull the container along. See how the elastic stretches. Measure it. The amount of stretch shows the amount of effort needed to pull the container.

❤ Experiment: Can the effort be changed?

See if you can change the effort needed. Do not change the container at all. Instead, change the surface under it. Make a track out of sandpaper and lay it along your desk. Hold the track in place with masking tape. Pull the container along it and measure the stretched elastic.

Picture 2   Set up the experiment like this

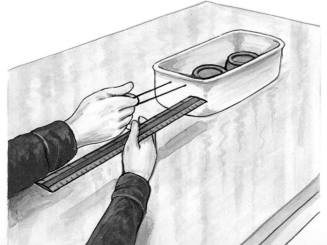
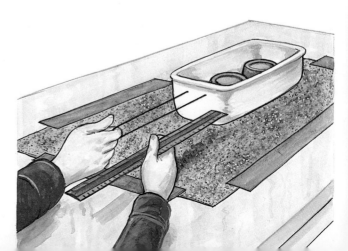

Repeat this experiment, using different tracks. You can make tracks out of cloth, cardboard, Plasticine, or anything handy. See how long the elastic stretches each time.

❤ Record

1 Make this chart to show your results:

What the track was made of?	The length of the elastic
along the desk	centimetres
along sandpaper	centimetres
along rubber	centimetres

2 Describe your experiments, and draw pictures to make things clear.
3 Say if it always takes an effort to drag things along.
4 Does the kind of surface affect the effort? What kind of surface makes the effort less? What kind of surface makes it greater?

Friction

Things drag when they are pulled past each other. We call this drag *friction*. Friction makes it hard to move things. Rough or sticky surfaces make most drag. The friction is greater. It was hardest to pull the log over the rough ground. Smooth surfaces make less drag. There is less friction. The smoother the surface, the less the effort needed to pull something. The log became easier and easier to pull as the pathway became smoother and smoother.

Friction helps

It is lucky for us that there is friction. Friction makes our feet grip the ground. It allows us to walk.

It is hard to walk where there is little friction. Think of walking on ice. It is so slippery that people fall over. Their feet cannot grip.

❤ To write

1 If there was no such thing as friction, could you walk?
2 Give an example of how friction can be made less.
3 How does friction help when you are riding a bicycle?

Picture 3   A bus on a skid-pan. This is good practise for the driver.

Unseen attraction

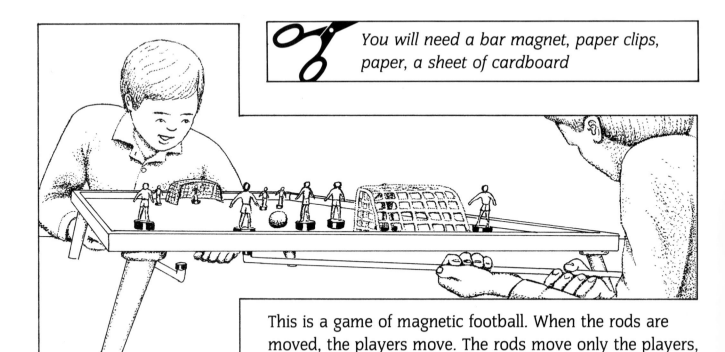

You will need a bar magnet, paper clips, paper, a sheet of cardboard

Picture 1 How does the game of magnetic football work?

This is a game of magnetic football. When the rods are moved, the players move. The rods move only the players, not the ball. How do the rods make the players move? Why do they not move the ball?

♥ An experiment to show what a magnet does

Put some paper clips on your desk. Hold a *magnet* near them. See how the magnet pulls the clips. Put the magnet near other things, and see if it pulls them. Make this chart to show what you find:

Pulled by the magnet	Not pulled by the magnet

A magnetic game

Take a sheet of cardboard and sprinkle paper clips over it. Then crumple up some little bits of paper into balls. Sprinkle these over the cardboard. Next, move the magnet underneath the cardboard. See which things move and which do not. See if you can move the paper clips off the cardboard without touching them.

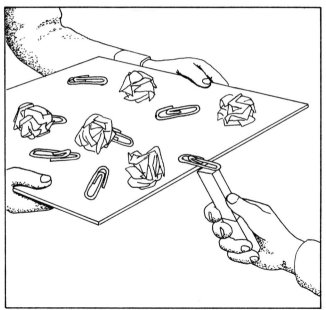

Picture 2 Finding out how a magnet works

❤ Record

1 Describe your experiments. Copy your chart into your book.
2 Say what your chart tells you about magnetism. Do magnets pull everything, or only some things? What are these things made of?
3 Did your magnetic game show you how magnetic football works? Say how you think it works.

Magnets

Men have known about magnets for a long time. Long ago, people discovered that some stones would pull iron. They called these stones *lodestones*. Lodestones were not made by men. They were made by nature.

Man-made magnets

Later, men found that iron itself could be made into magnets. These magnets were stronger than lodestones. Later still, men invented the metal which we call steel. Steel can be turned into magnets which are even better than iron ones. Modern magnets are made of iron or steel.

Magnetism at work

Magnets pull metals, but not other things. There are some kinds of metal that magnets will not attract.

Magnetism works even when other things are put in the way. For example, your magnet was able to pull the paper clips even through the cardboard.

Magnetic football

Here is how the game works. There are magnets at the ends of the rods. Each little footballer has metal underneath. The magnets pull this metal, and the players move. The ball is made of plastic. Magnets do not pull plastic, so the rods do not make the ball move.

❤ To write

1 Which of these things will a magnet pick up: pins; peanuts; nails; Smarties; a small knife; a pair of glasses.
2 What are magnets made of?
3 What are lodestones?

Picture 3 This car aerial is held on by a magnet

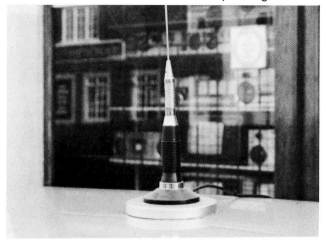

Things we cannot see

Picture 1   Gravity pulls these parachutists downward

Picture 2   Why do apples fall to the ground?

Forces

The world is full of things that we can see, such as soil, plants, animals and clouds. There are things in the world that we cannot see. We say that they are invisible. We call these invisible things *forces*. Some facts about these forces were known long ago. Some have been discovered in modern times.

Gravity

Early men knew about gravity. They knew that if they dropped something, it would fall. They did not know what caused gravity.

In more modern times, people thought about gravity. The most famous of these people was Isaac Newton. He wrote about gravity in 1687. He decided that the pull of gravity was caused by the Earth.

The Earth causes this force because it is so big and solid. It is massive. Isaac Newton said that all massive things pull other things.

The moon has its own gravity. It pulls things to itself. It is massive enough to do this. The moon is made of solid rock, and it is big, but the gravity of the moon is weaker than the gravity of Earth.

Friction

Friction is invisible, but it makes things happen. It makes it hard to slide one thing over another. It can also make heat. If you rub your hand hard on the desk, it will get hot. Early people discovered this and found a way to use it. They twirled pieces of wood together. There was friction between the pieces of wood. They grew hot. After a time, they grew hot enough to start a fire.

Magnetism

Magnetism was known long ago. In about 420 AD, a man called Augustine described a lodestone. A lodestone is a natural magnet. He described how a lodestone under a tray could move a piece of metal on top.

People were using magnets in England over eight hundred years ago. They used them to help them find their way, like we use compasses nowadays.

Over seven hundred years ago, a Frenchman experimented with lodestones. His name was Petrus Peregrinus de Maricourt. He found that parts of a lodestone were more magnetic than other parts. He called the stronger parts the *poles* of the lodestone. We still call the strongest parts of magnets the poles.

Picture 3 Using friction to start a fire

Picture 4 We use magnets today for different jobs. This container is being lifted by a huge magnet.

New words

atmosphere   the layer of air that surrounds the Earth

earthquake   a shaking of the surface of the Earth
Earth's crust   a thick layer of rock that lies under the soil
experiment   a test designed to find out more about something

force   something invisible that can affect the way that things behave
friction   the drag that happens when two things slide past each other

globe   something with the shape of a ball
gravity   the force that pulls things down towards the centre of the Earth

inorganic   not made of living, or dead, plants or animals

lodestone   a kind of rock that has natural magnetism

magma   molten rock inside the Earth
magnet   something that attracts certain kinds of metal
magnetism   the invisible pull that a magnet has
mass   the amount of stuff that there is in anything
massive   made of a lot of solid stuff
minerals   inorganic things like water and stone
molten   melted
moons   rocky globes that circle around planets

organic   living, or dead, plants or animals
organism   a living thing

planets   large globes, made mostly of rock, that circle around stars
poles   the strongest parts of a magnet

science   the study of why things happen
scientist   a person who studies science
space   the huge emptiness in which all stars and planets float
stars   suns that are so far away that they only look like points of light
subsoil   soil that lies a little below the surface and is not good for growing things

topsoil   the thin layer of soil at the top of the ground that is good for growing things

weightlessness   where there is no weight because there is no gravity

zero gravity   has the same meaning as *weightlessness*